MadTest

GO NO GO Series on Secondary Injection Testing,

Model ABB, Gould BBC, Brown Boveri, ITE

504 Power Shield Test Kit

Book 1

1st Edition

Joseph D. Paresi

Madison, Connecticut USA

2018

Description

GO-NO-GO procedures for the 504 Power Shield Test Kit.

An easy to use GO-NO-GO procedural guide on the correct operation of the 504 Power Shield Test Kit. In the series we use our own universal format to explain the use of each test kit. The procedures are designed to simply verify the functionality of the test set while testing a power shield 504 trip unit. The use of this book is in no way a substitute for calibration procedures on the test set or trip unit programmer. The author's intention is simply to compliment the OEM's instruction bulletin that was provided with the test set. The GO-NO-GO procedures were originally created to help with testing in between test equipment rentals or use periods. The idea was simply to physically demonstrate an actual test by means of photographs and performing the testing while displaying the authors results. Another important aspect of this series was to provide a history of the test set itself along with a brief time line of the product line. We are excited to offer this GO-NO-GO procedure to the electrical testing world an offer an alternative means of educating the end user of the 504 Power Shield test kit.

COMPONENT CHECK LIST FOR

GO-NO-GO PROCEDURES FOR

GOULD-BROWN BOVERI-ABB 504 TEST SET

Components and accessories check list;

The BROWN BOVERI / GOULD / ABB 504 Trip Test Set is composed of five components listed below;

See Picture A. page 5

1. BROWN BAVOVERI/ GOULD/ ABB Test Set Model 504 **ABB Part # 608-089T1**
2. POWER SHIELD INTERFACE CABLE **ABB Part # 608-082-T1**
3. SENSOR CHECK TEST LEAD
4. POWER CABLE 120V
5. IB-9.1.7-22 Issue B. Instruction Bulletin

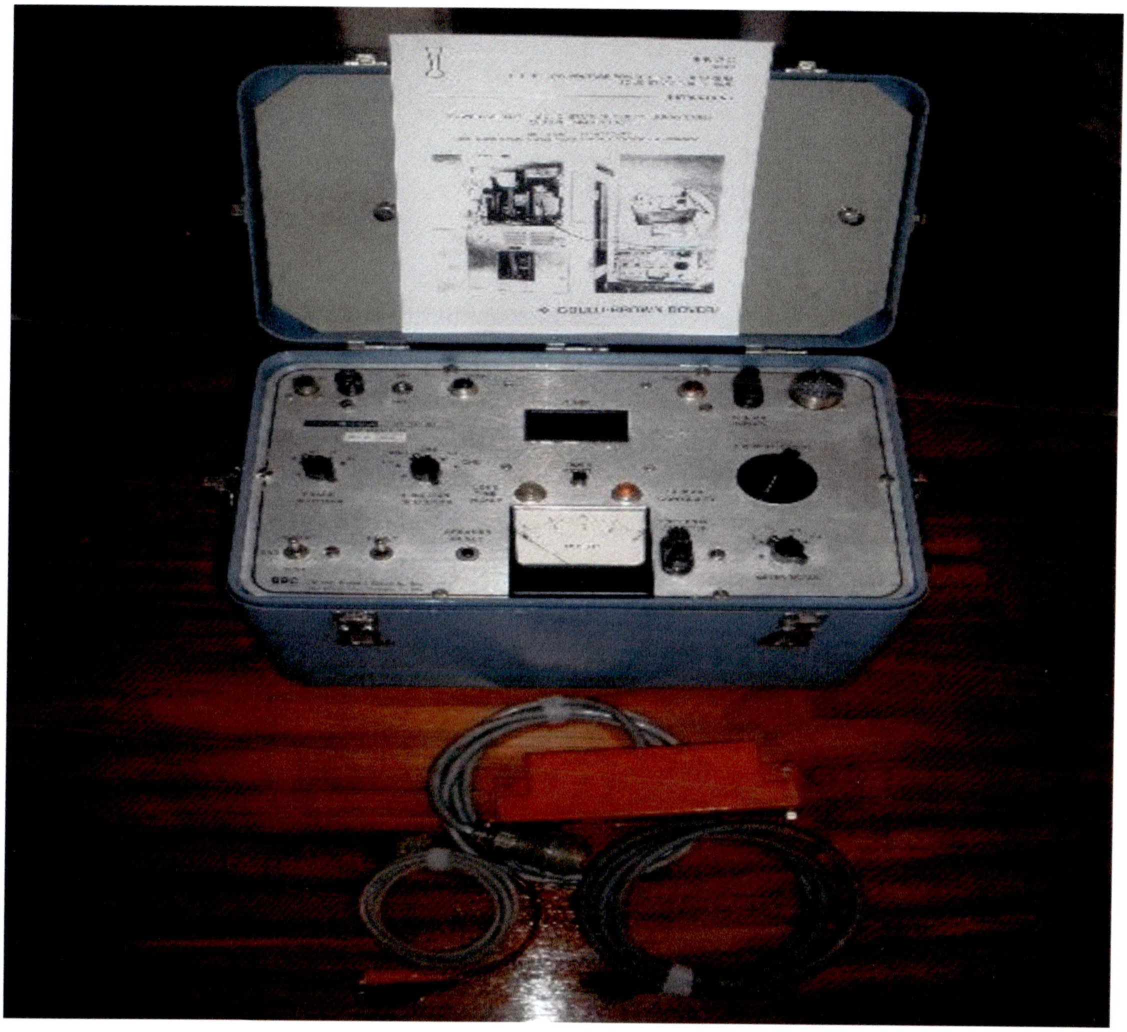

A.

In the photo above is the BROWN BOVERI/GOULD/ABB Test Set Model 504 shown with factory supplied Power Shield interface cable, Sensor test lead, Power cable, with IB-9.1.7-22 Issue B instruction bulletin. The **GO-NO-GO** procedures were developed in part with a combination of the manufactures IB9.1.1-22 Instructional Bulletin, as well as actual hands on verification with most of the functions that the **GOULD-BROWN BOVERI/GOULD/ABB 504 Test Set** can test. The goal in developing this **GO-NO-GO** procedure was to clarify and simplify the use of the test set and to allow a technician to quickly go through the steps of the **GO-NO-GO** procedural guide and verify that the test set is functioning properly function in the field or in between use or rental periods.

BRIEF HISTORY OF THE TEST KIT AND PRODUCT LINE

The GOULD-BROWN BOVERI –ABB 504 Test set is known in the service industry as the "**ABB-504 Test Set**". The reason for this is over time the acquisitions of the following; ITE-Circuit Breaker Company named for the innovative oil-dashpot inverse time element on its electro-mechanical circuit breaker trip devices, was acquired in the late seventies by Gould, Inc. The new company Gould ITE continued to manufacture its products mainly in several factories in the Philadelphia area and a west coast facility in Downey, CA. Gould sold a majority of the ITE Company to Swiss based Brown Boveri Limited, which later became **ABB**. The POWER SHIELD Trip Programmer which is the programmer that the ABB-504 tests sets are currently still in production and is supported by ABB. For **GO-NO-GO** explanations we will simply refer to the test set as the **ABB 504 Test Set**.

Shown in the photo below the POWER SHIELD Solid State Trip Type SS5G programmer is supplied to verify the **GO-NO-GO** procedures. The unit was removed from a **K-1600S ITE** Circuit Breaker. Before any testing can begin with the ABB-504 Test Set **you must know** the amperage rating of the Circuit Breaker it was removed from. The trip units have specific internally installed trip settings for each circuit breaker.

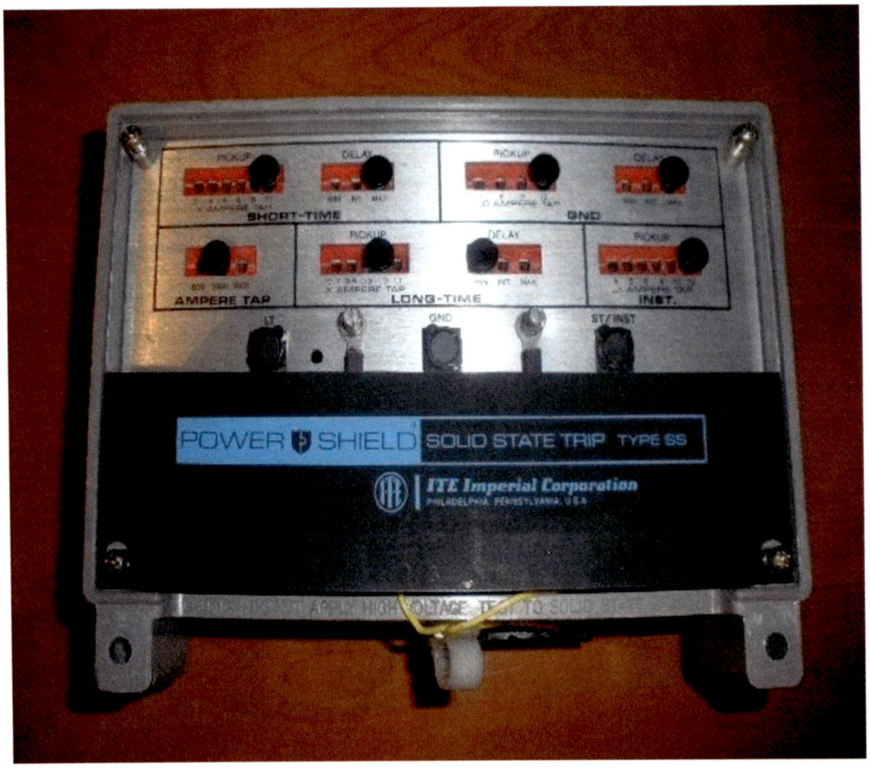

B.

C.

Shown in the above photo the cover plate removed from the Power Shield SS 5G Programmer.

The ABB-504 test set has specifically been designed to use with the POWER SHIELD programmer and incorporates most of the required test circuitry in a compact test set. The **GO-NO-GO** procedures shall be used in conjunction with the IB-9.1.7-22 instruction bulletin. The following tests will allow a technician to check the following; **Long Time, Short Time, Instantaneous, and Ground functions.**

INDEX

Preliminary connection from ABB 504 to Programmer Trip Unit.

D.

E.

In the pictures above, you are looking at the programmer and interphase connection with the cover plate removed. The ABB inter phase cable needs to attach with the two tabs that fit into the programmer underneath the face plate. Since the adapter is spring loaded you must gently apply pressure while connecting the two components. While applying pressure the adapter is in the lowest position you then tighten the thumb screws shown in picture E. You are then connected to your programmer and ready for connection to the test set.

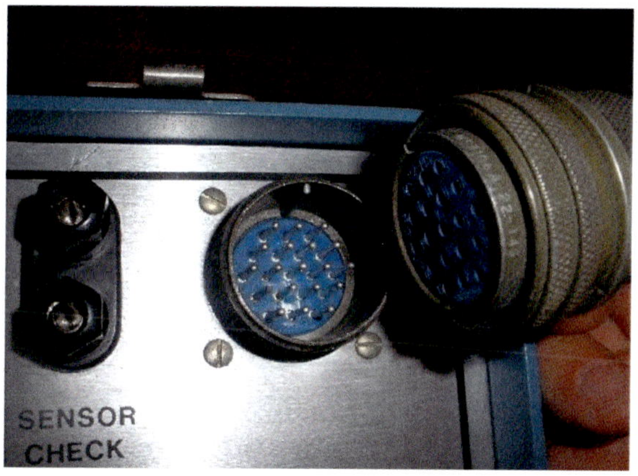

F.

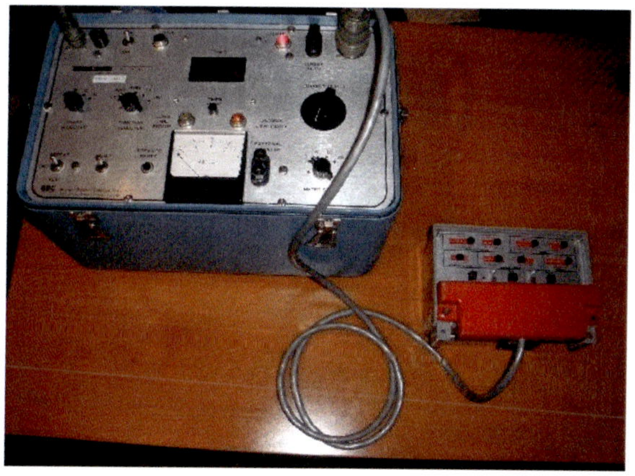

G.

In the above left picture, the connector to the test set, make sure the upper slot matches up with the both the male and female half, completely tighten down the connector. In picture G. above right the ABB 504 and Power Shield SS5G Programmer are ready to begin testing. Be aware the red insulation material at the end of the cable harness can become brittle over time and the tabs that hold the tight down screws break off. This will result in a repair or a modification to use the connector properly. Careful attention should be addressed while connecting to the Power Sheild trip unit.

BEFORE YOU BEGIN

Before testing can begin there are 5 steps that are covered in the manual to insure you are aware of the test set preliminary settings. For **GO-NO-GO** procedures they will also be used.

NOTE 1

When testing the "POWER SHIELD" programmer units equipped with ground protection it will be necessary to set the GND switch in the "DEFEAT" position. The "POWER SHIELD" SS 5G programmer has been deliberately selected with this feature. **See Picture H, page 11**. This is only for the Long-Time, Instantaneous, and Short-Time delay functions. For Ground Trip functions position the switch back to the "TEST" position.

NOTE 2

The "CLOSE" lamp must be lighted to begin any test. If not, depress the "RESET" button. **See Picture I, page 11.**

NOTE 3

For field applications the "POWER SHIELD" functions can be tested with the circuit breaker in the open position. When this method is performed the TRIP and CLOSE lights will simulate breaker operation.

NOTE 4

Tolerances of ± 15% ***are suggested*** to account for field testing variations. For GO-NO-GO procedures this same tolerance shall be accepted.

NOTE 5

For **GO-NO-GO** procedures the POWER SHEILD programmer had two resistor leads that must be connected for testing off the Circuit Breaker. The terminal leads are 3 and 16 and are intentionally connected, **See Picture J, page 11.**

H. **I.**

In the picture above left the DEFEAT position is selected for LT, ST, and INST testing, when Ground Trip functions you set it to TEST. Above right the CLOSE lamp MUST be on at the beginning of any test.

J.

The above picture the yellow wires are connected to terminals 3 and 16 to a supplied resistor. For GO-NO-GO procedures this is an intentional connection.

LONG TIME FUNCTION TEST

PURPOSE

Verify that pickup occurs within tolerance ± 15%. With the GO-NO-GO procedures the programmer there is a convenient indicating LONG TIME PICKUP light permitting a straight forward visual means of determining when the LONG TIME PICKUP point is reached. **See Picture 3.A, page 13.**

GO-NO-GO

LT PICKUP TEST

Test Procedure

1. Programmer Settings:

   Insure all pins that are not being tested are at their maximum or highest settings.

   Set pin in Long Time Pickup position 0.7, **See Picture 1.A, page 13.**

   Set pin in the 1600 Ampere Tap, **See Picture 2.A, page 13.**

   Set pin in Long Time Delay Setting to MIN, **See Picture 4.A, page 14.**

2. Position Test Set Controls:

   SET the Function Selector to LT, **See Picture 5.A, page 14.**

   SET the Phase Selector to desired phase A, B or C, **See Picture 6.A, page 14.**

   Set the DEFEAT, GROUND, TEST switch to DEFEAT, **See Picture 7.A, page 14.**

   Press the Timer Reset button, **See Picture 8.A, page 14.**

   Press the Breaker Reset button, **See Picture 9.A, page 14.**

   Current adjust should be in the zero position.

   Note the close breaker light should on**, See Picture 8.A, page 14.**

3. Position Test Set Controls you can now begin to test:

   To begin hold down the fault toggle switch and hold it with your left hand, **Picture 3.A page 13.** Slowly raise the current adjust until you Long Time Pick-Up light lights up. The light should light up when the **1.4 amperes** mark is made on the 5-ammeter scale, stop at that point, ramp down the current adjust control and you can leave the timer running. Move the SET Long Time Pickup Pin to 0.9 repeat the test the light should light up to 1.8 **amperes.** Again, ramp down the current adjust to zero and move the pin to 1.0 repeat the test and the light should light up at **2 amperes** on the 5-ammeter scale. You can leave the timer running and if desired you can check A, B and C phase the same results should occur. For **GO-NO-GO** purposes pick any phase and start the test one phase was determined to verify if the test set was working correctly. The calculations are based on **Table 2** of **IB-9.1.7-22** manual, **Picture 11.A and 12.A, page 15.**

1. A

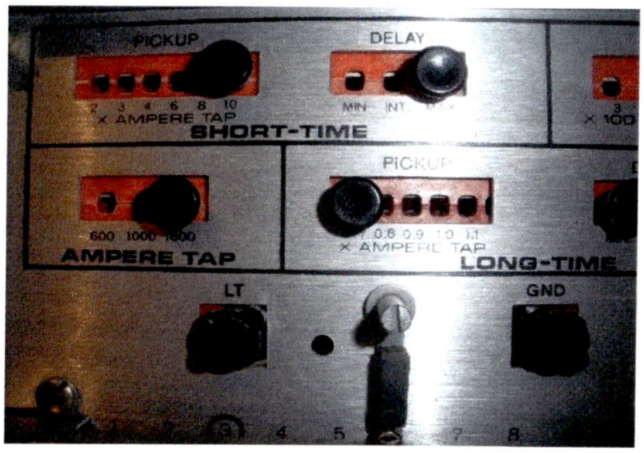

2. A

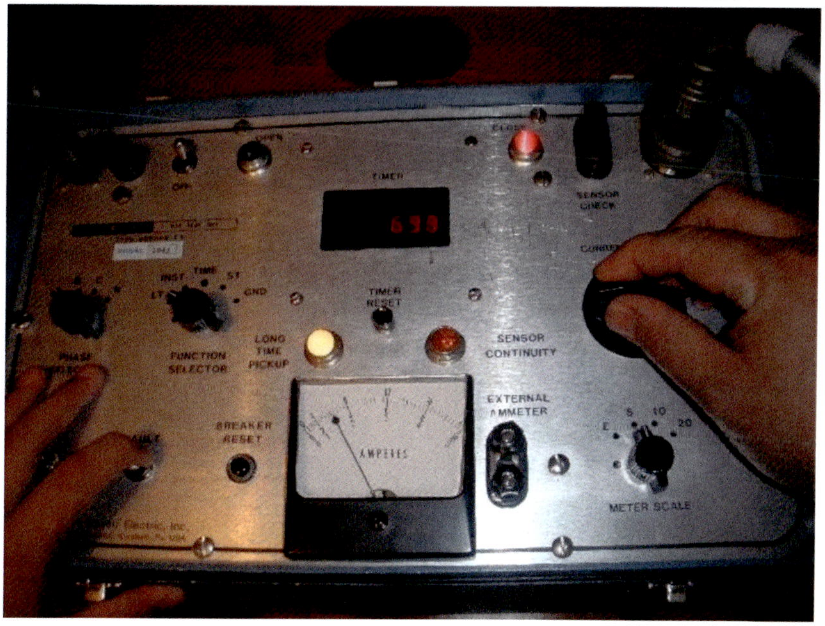

3.A

4. A

5. A

6. A

7. A

8. A

9. A

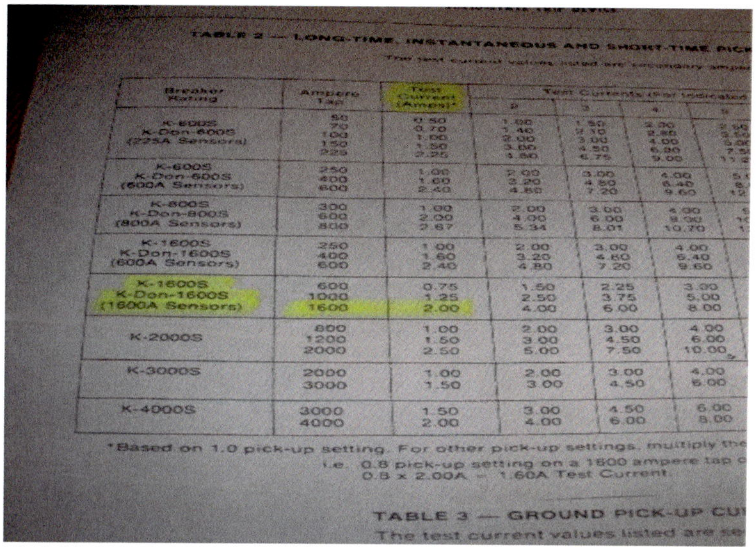

TABLE 2 — LONG-TIME, INSTANTANEOUS AND SHORT-TIME PICK

The test current values listed are secondary ampe

Breaker Rating	Ampere Tap	Test Current (Amps)*	Test Currents (For Indicated 2	3	4
K-600S K-Don-600S (225A Sensors)	50 70 100 150 225	0.50 0.70 1.00 1.50 2.25	1.00 1.40 2.00 3.00 4.50	1.50 2.10 3.00 4.50 6.75	2.00 2.80 4.00 6.00 9.00
K-600S K-Don-600S (600A Sensors)	250 400 600	1.00 1.60 2.40	2.00 3.20 4.80	3.00 4.80 7.20	4.00 6.40 9.60
K-800S K-Don-800S (800A Sensors)	300 600 800	1.00 2.00 2.67	2.00 4.00 5.34	3.00 6.00 8.01	4.00 8.00 10.70
K-1600S K-Don-1600S (600A Sensors)	250 400 600	1.00 1.60 2.40	2.00 3.20 4.80	3.00 4.80 7.20	4.00 6.40 9.60
K-1600S K-Don-1600S (1600A Sensors)	600 1000 1600	0.75 1.25 2.00	1.50 2.50 4.00	2.25 3.75 6.00	3.00 5.00 8.00
K-2000S	800 1200 2000	1.00 1.50 2.50	2.00 3.00 5.00	3.00 4.50 7.50	4.00 6.00 10.00
K-3000S	2000 3000	1.00 1.50	2.00 3.00	3.00 4.50	4.00 6.00
K-4000S	3000 4000	1.50 2.00	3.00 4.00	4.50 6.00	6.00 8.00

*Based on 1.0 pick-up setting. For other pick-up settings, multiply the

i.e. 0.8 pick-up setting on a 1600 ampere tap o

0.8 x 2.00A = 1.60A Test Current.

TABLE 3 — GROUND PICK-UP CU

The test current values listed are se

11.A

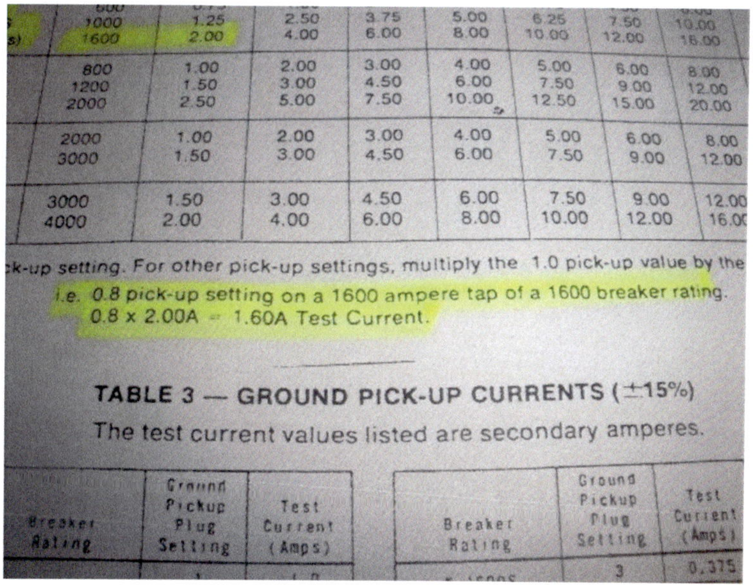

1000 1600	1.25 2.00	2.50 4.00	3.75 6.00	5.00 8.00	6.25 10.00	7.50 12.00	10.00 16.00
800 1200 2000	1.00 1.50 2.50	2.00 3.00 5.00	3.00 4.50 7.50	4.00 6.00 10.00	5.00 7.50 12.50	6.00 9.00 15.00	8.00 12.00 20.00
2000 3000	1.00 1.50	2.00 3.00	3.00 4.50	4.00 6.00	5.00 7.50	6.00 9.00	8.00 12.00
3000 4000	1.50 2.00	3.00 4.00	4.50 6.00	6.00 8.00	7.50 10.00	9.00 12.00	12.00 16.00

ck-up setting. For other pick-up settings, multiply the 1.0 pick-up value by the

i.e. 0.8 pick-up setting on a 1600 ampere tap of a 1600 breaker rating.

0.8 x 2.00A = 1.60A Test Current.

TABLE 3 — GROUND PICK-UP CURRENTS (±15%)

The test current values listed are secondary amperes.

Breaker Rating	Ground Pickup Plug Setting	Test Current (Amps)	Breaker Rating	Ground Pickup Plug Setting	Test Current (Amps)
				3	0.375

12.A

In the photos above, chart calculations are based on the programmer type K-1600S K-Don 1600S with 1600-amp sensors. The areas in yellow have been highlighted so clearly show where the current calculations come from shown in above **picture 11.A**. For correct testing the type of circuit breaker the programmer is removed from is essential to appropriately use the above charts located in the OEM IB-9.1.7-22. The GO-NO-GO procedures are based on the tap on 1600 and shown above. The multiplier example is shown above **picture 12.A**

LONG TIME FUNCTION TEST (CONT)
PURPOSE

Verify that pickup occurs within tolerance ± 15%. This test is for **delay pickup** calculations based on **3 X** of any of the three pickup settings from the Pick-Up tests. All calculations are based on TABLE 4 –Trip Times, in the **IB-9.1.7.22** manual page 9. For **GO-NO-GO** procedures and long trip results only 3 tests are used.

GO-NO-GO

LT DELAY TEST

Test Procedure

1. Programmer Settings:

   Insure all pins that are not being tested are at their maximum or highest settings.

   Set pin in and leave the Long Time Pickup position 0.7, **See Picture 1.A, page 13.**

   Set pin in the 1600 Ampere Tap, **See Picture 2.A, page 13.**

   Set pin in Long Time Delay Setting to MIN, **See Picture 4.A, page 14.**

2. Position Test Set Controls:

   SET the Function Selector to TIME, **See Picture 1.B, page 17.**

   SET the Phase Selector to desired phase A is fine, **See Picture 2.B, page 17.**

   Keep the switch position of DEFEAT, GROUND, TEST to DEFEAT, **See Picture 7.A page 14.**

   Set the ammeter scale to 5, **See Picture 3.B, page 17.**

   Preset current to the desired preset amps, based on table 4, **See Picture 4.B, page 17.**

   Press the Timer Reset button, **See Picture 5.B, page 17.**

   Press the Breaker Reset button, **See Picture 6.B, page 17.**

   Note: The close breaker light should be on, **See Picture 4.B, page 17.**

(Testing Continued) on page 19.

1.B

2.B

3.B

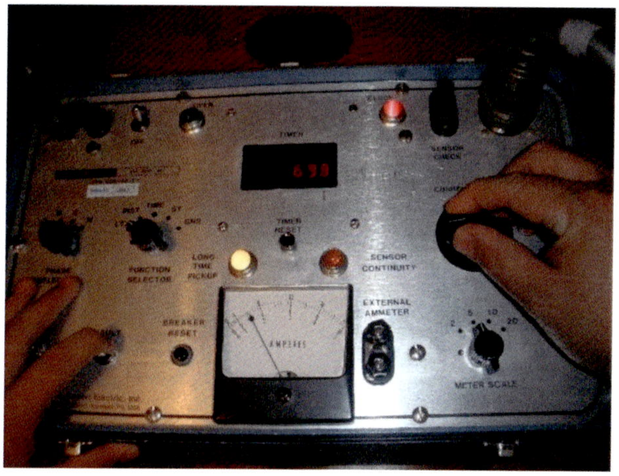

4.B

5.B

6.B

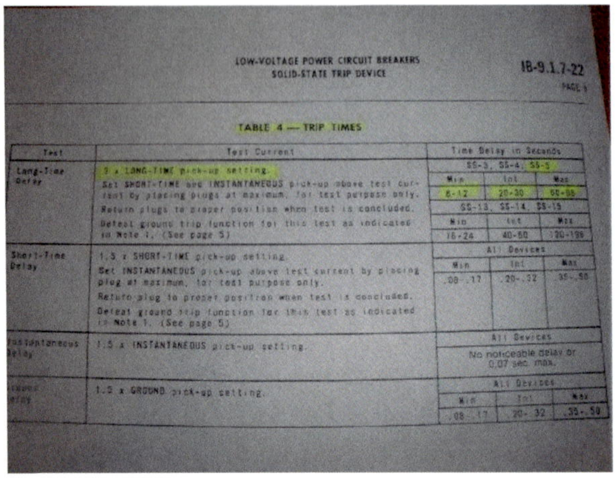

7.B

8.B

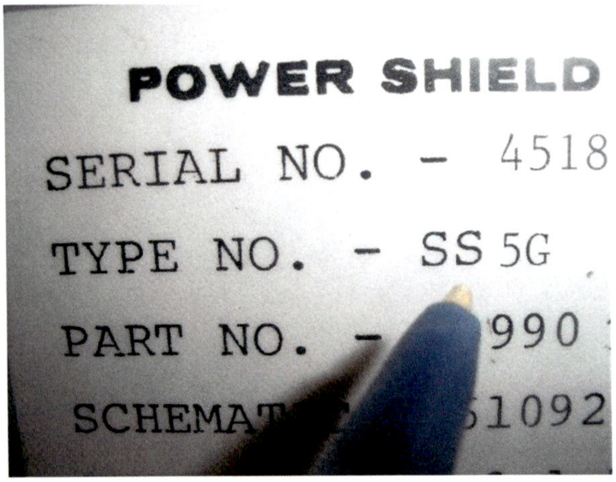

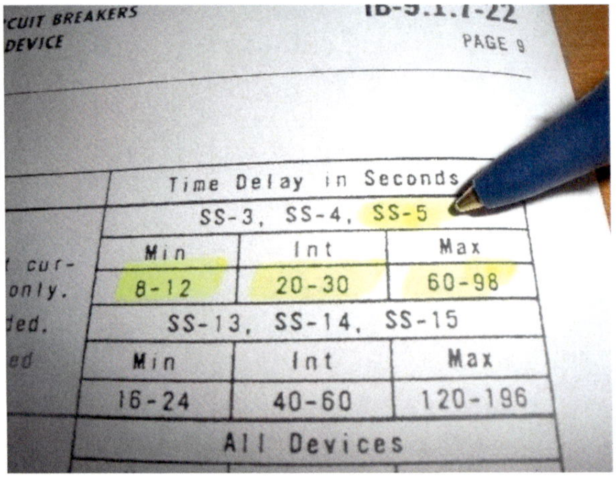

9.B

10.B

11.B

12.B

3. You can now begin to test:

You must first calculate the pre-set current **.7 X 2 X 3 Times** = or .7 (X Long Time Amperes) X 2 (Based on Test Current Amps Table 2) X 3 Times Long Time pickup setting = **4.2 amps,** preset the current to the desired amps, **See Picture 4.B, page 17.**

Note: The Power Shield Programmer unit is a TYPE NO. SS 5G which determines your Time Delay Set Times, **See Picture 9.B, page 18**, and illustrated in yellow highlight in the table 4, **See Picture 10.B, page 18.**

Hold down the Fault button and if needed adjust the current to the desired 4.2 amperes. The trip unit will trip under 12 seconds.

Reset the following;

Timer Reset, **See Picture 5.B, page 17.**

Breaker Reset, **See Picture 6.B, page 17.**

Change Phase Select, to Phase B, then C, or A.

Change Delay Pin to INST, then for the next test MAX shown on chart below.

On the programmer rotate the magnet on the LT target to the 9 O'clock position and set it back to the 6 O'clock position that will set the target back to black, **See Picture 12.B, page 16.**
NOTE: You can use your fingers to adjust the magnet or use a small adjustment screw driver.

Repeat the test, see below results.

Delay	**Phase**	**Preset Amperes**	**GO-NO-GO Results**	**Table 4 Set Times**
MIN	A.	4.2 Amperes	11.81 Seconds	8-12 Seconds
INST	B.	4.2 Amperes	29.01 Seconds	20-30 Seconds
MAX	C.	4.2 Amperes	86.84 Seconds	60-98 Seconds

The table above was created from random test results based on the authors actual testing, the end user should experience similar results, the author found 3 tests were sufficient for GO-NO- GO.

INSTANTANEOUS FUNCTION TEST

PURPOSE

Verify that pickup occurs within tolerance ± 15%. For the following **GO-NO-GO** procedures the programmer's target will indicate a trip when the calculated setting is reached, **See Picture 1.C, page 21.**

GO-NO-GO

PICKUP TEST

Test Procedure

1. Programmer Settings:

   Insure all pins that are not being tested are at their maximum or highest settings.

   Set Instantaneous Pin to 4 X position, **See Picture 2.C, page 21.**

2. Position Test Set Controls:

   SET the Function Selector to INST, **See Picture 3.C, page 21.**

   SET the Phase Selector to desired phase A, **See Picture 4.C, page 21.**

   Set the DEFEAT, GROUND, TEST switch to DEFEAT, **See Picture 5.C, page 21.**

   Current adjust should be in the zero position.

   Press the Timer Reset button, **See Picture 5.B, page 17.**

   Press the Breaker Reset button, **See Picture 6.B, page 17.**

   Note the close breaker light should on, **See Picture 8.A, page 14.**

(Testing Continued) on page 22.

1.C

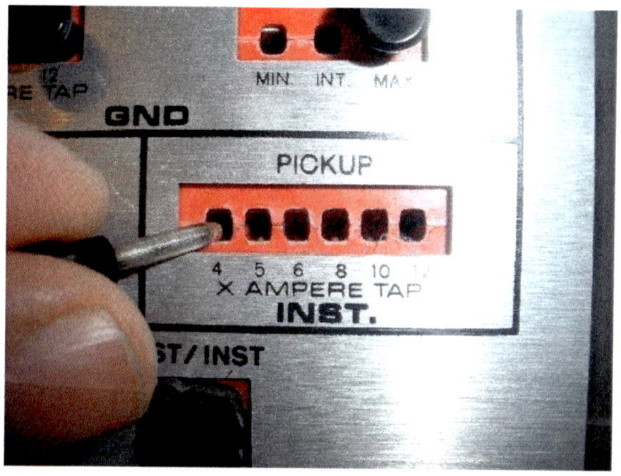

2.C

3.C

4.C

5.C

6.C

3. You can now begin to test:

   You must first calculate the preset current **4 X 2 = 8 amps.** = or 4 (X INST Amperes Tap) X 2 = 8 amps, (Based on Test Current Amps Table 2), **See Picture 11.A, page 17.**

   CHANGE the meter scale to 10, then 20 for the last test, **See Picture 3.B, page 19.**

   Hold down the Fault button and adjust the current to the desired 8 amperes. The programmer will trip at 8 amps or ± 15% of that number.

   Reset the following;

   Timer Reset, See **Picture 5.B, page 19.**

   Breaker Reset, **See Picture 6.B, page 19.**

   If desired you can change Phase Select, to Phase B or C.

   Change INST Pin to 6, then for the next test to pin 10 shown on chart below.

   On the programmer rotate the magnet on the **INST target** to the 3 O'clock position and set it back to the 6 O'clock position that will set the target back to black, **See Picture 6.C, page 23.**

INST AMPERE TAP	**Phase**	**Calculated Amperes**	**GO-NO-GO Trip Results**
4	A.	4 Amperes	4 Amperes
6	A.	12 Amperes	12 Amperes
10	A.	20 Amperes	20 Amperes

The table above was created from test results based on the authors actual testing, the lab should experience similar results, the author found 3 tests were sufficient for GO-NO-GO.

SHORT TIME FUNCTION TEST

PURPOSE

Verify that pickup occurs within tolerance ± 15%. For the following **GO-NO-GO** procedures the programmer's target will indicate a trip when the calculated setting is reached, **See Picture 1.C, page 21.**

GO-NO-GO

PICKUP-UP TEST

Test Procedure

1. Programmer Settings:

   Insure all pins that are not being tested are at their maximum or highest settings.

   Set pin in Short Time Pickup position 2, **See Picture 1.D, page 24.**

   Set pin in the Short Time Delay pin to MIN Tap, **See Picture 2.D, page 24.**

2. Position Test Set Controls:

   SET the Function Selector to ST, **See Picture 3.D, page 24.**

   SET the Phase Selector to desired phase B, **See Picture 4.D, page 24.**

   Set the DEFEAT, GROUND, TEST to DEFEAT, **See Picture 5.C, page 21.**

   Press the Timer Reset button, **See Picture 5.B, page 17.**

   Press the Breaker Reset button, **See Picture 6.B, page 17.**

   Current adjust should be in the zero position.

   Note the close breaker light should on**, See Picture 8.A, page 14.**

(Testing Continued) on page 25.

1.D

2.D

3.D

4.D

5.D

6.D

3. You can now begin to test:

   You must first calculate the preset current **2 X 2 = 4 amps.** = or 2 (X INST Amperes Tap) X 2 = 4 amps, (Based on Test Current Amps Table 2), **See Picture 11.A, page 15.**

   CHANGE the meter scale between tests to **5**, **10**, and then **20** for the last test, **See Picture 5.D, page 24.**

   Hold down the Fault button and adjust the current to the desired 4 amperes. The programmer will trip at 4 amps or ± 15% of that number.

   Reset the following;

   Timer Reset, **See Picture 5.B, page 17.**

   Breaker Reset, **See Picture 6.B, page 17.**

   If desired you can change Phase Select, to A or C Phase.

   Change INST Pin to 4, then for the next test to pin 8 shown on chart below.

   On the programmer rotate the magnet on the **INST target** to the 3 O'clock position and set it back to the 6 O'clock position that will set the target back to black, **See Picture 6.C, page 21.**

SHORT-TIME AMPERE TAP	**Phase**	**Calculated Amperes**	**GO-NO-GO Trip Results**
2	B.	4 Amperes	4 Amperes
4	B.	8 Amperes	8 Amperes
8	B.	16 Amperes	16 Amperes

The table above was created from test results and based on the authors actual testing, the lab should experience similar results, the author found 3 tests were sufficient for GO-NO-GO.

SHORT TIME FUNCTION TESTS (CONT)

PURPOSE

Verify that pickup occurs within tolerance ± 15%. This test is for **delay pickup** calculations based on 150% of pickup of any of the three pickup settings from the Pick-Up tests. All calculations are based on TABLE 4 –Trip Times for Short-Time Delay, in the **OEM IB-9.1.7.22** manual page 9. For **GO-NO-GO** procedures and long trip results only 3 tests are used.

GO-NO-GO

ST DELAY TEST

Test Procedure

1. Programmer Settings:

   Insure all pins that are not being tested are at their maximum or highest settings.

   Set the Short Time Pickup pin to pick-up position 10, **See Picture 2.E, page 27.**

   Set pin in Short Time Delay Setting to MIN.

2. Position Test Set Controls:

   SET the Function Selector to TIME, **See Picture 3.E, page 27.**

   SET the Phase Selector to desired phase A.

   Keep the position of DEFEAT, GROUND, TEST to DEFEAT.

   Set the ammeter scale to 10, **See Picture 4.E, page 27.**

   Preset current to the desired preset amps, based on table 4, **See Picture 1.E, page 27.**

   The pickup delay setting should be at 150% of pickup.

   Press the Timer Reset button.

   Press the Breaker Reset button.

   Note: The close breaker light should be on.

(Testing Continued) on page 28.

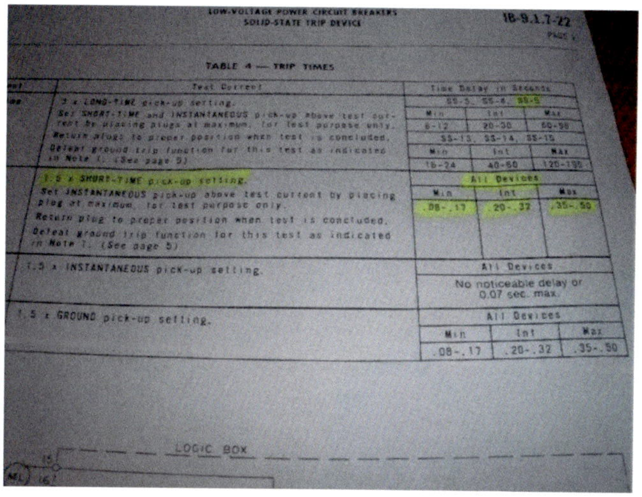
LOW-VOLTAGE POWER CIRCUIT BREAKERS
SOLID-STATE TRIP DEVICE
IB-9.1.7-22

TABLE 4 — TRIP TIMES

Test Current	Time Delay in Seconds		
3 x LONG-TIME pick-up setting. Set SHORT-TIME and INSTANTANEOUS pick-up above test current by placing plugs at maximum, for test purpose only. Return plugs to proper position when test is concluded. Defeat ground trip function for this test as indicated in Note 1. (See page 5)	SS-5, SS-6, SS-9		
	Min	Int	Max
	6-12	20-30	60-90
	SS-13, SS-14, SS-15		
	Min	Int	Max
	16-24	40-60	120-180
1.5 x SHORT-TIME pick-up setting. Set INSTANTANEOUS pick-up above test current by placing plug at maximum, for test purpose only. Return plug to proper position when test is concluded. Defeat ground trip function for this test as indicated in Note 1. (See page 5)	All Devices		
	Min	Int	Max
	.08-.17	.20-.32	.35-.50
1.5 x INSTANTANEOUS pick-up setting.	All Devices		
	No noticeable delay or 0.07 sec. max.		
1.5 x GROUND pick-up setting.	All Devices		
	Min	Int	Max
	.08-.17	.20-.32	.35-.50

LOGIC BOX

1.E

2.E

3.E

4.E

5.E

Before you can begin your test, you must first calculate the preset current **2 X 2 x 150% = 6 amps.** = or 2 (X SHORT-TIME Amperes Tap) X 150% = 6 amps, (Based on Test Current Amps Table 4), **See Picture 1.E, page 27.**

Preset your current to 6 amps and leave the current adjust in position. Move the SHORT- TIME Ampere tap to the #10 position.

After you preset your current remove the SHORT-TIME Ampere Tap back to 2.

3. You can now begin to test:

   Hold down the Fault Button. The programmer will trip within the time table in Table 4.

   Reset the following;

   Timer Reset, **See Picture 5.B, page 17.**

   Breaker Reset, **See Picture 6.B, page 17.**

   If desired you can change Phase Select, to A, B or C Phase.

   Change DELAY Pin to MIN pin, after test to INST pin, then to next test to MAX pin shown on chart below.

   On the programmer between each test, rotate the magnet on the **INST target** to the 3 O'clock position and set it back to the 6 O'clock position that will set the target back to black, **See Picture 6.C, page 21.**

Delay	**Phase**	**Preset Amperes**	**GO-NO-GO Results**	**Table 4 Set Times**
MIN	A.	6 Amperes	.11 Seconds	.08-.17 Seconds
INST	B.	6 Amperes	.23 Seconds	.20-.32 Seconds
MAX	C.	6 Amperes	.41 Seconds	.35-.50 Seconds

The table above was created from test results based on the authors actual testing, the lab should experience similar results, the author found 3 tests were sufficient for GO-NO-GO.

GROUND FAULT FUNCTION TEST

PURPOSE

Verify that pickup occurs within tolerance ± 15%. For the following **GO-NO-GO** procedures the programmer's ground target will indicate a trip when the calculated setting is reached, **See Picture 1.C, page 21.**

GO-NO-GO

GF PICKUP TEST

Test Procedure

1. Programmer Settings:

   Insure all pins that are not being tested are at their maximum or highest settings.

   Set pin in Ground Pickup tap to position 0.3, **See Picture 1.F, page 30.**

   Set pin in the Short Time Delay pin to MAX Tap, **See Picture 2.F, page 30.**

2. Position Test Set Controls:

   SET the Function Selector to GND, **See Picture 3.F, page 30.**

   SET the Phase Selector to desired phase A, B, or C.

   Set the DEFEAT, GROUND, TEST switch to TEST, **See Picture 4.F, page 30.**

   Press the Timer Reset button.

   Press the Breaker Reset button.

   Current adjust should be in the zero position.

   Note the close breaker light should on.

(Testing Continued) on page 31.

1.F

2.F

3.F

4.F

5.F

6.F

3. You can now begin to test:

   Hold down the Fault Switch and slowly adjust the Current Adjust Control until the Open Lamp lights and the programmer unit trips. The programmer will trip within the time table in Table 3. The settings are based on **Table 3** of **IB-9.1.7-22** manual, **See Picture 7.F, page 32.** Reset the following;

   Timer Reset.

   Breaker Reset.

   If desired you can change Phase Select to A, B, or C Phase.

   Change INST pin to 6, then 9, and 12 between tests shown on chart below.

   On the programmer rotate the magnet on the **GRD target** to the 3 O'clock position and set it back to the 6 O'clock position that will set the target back to black, **See Picture 6.F, page 30.**

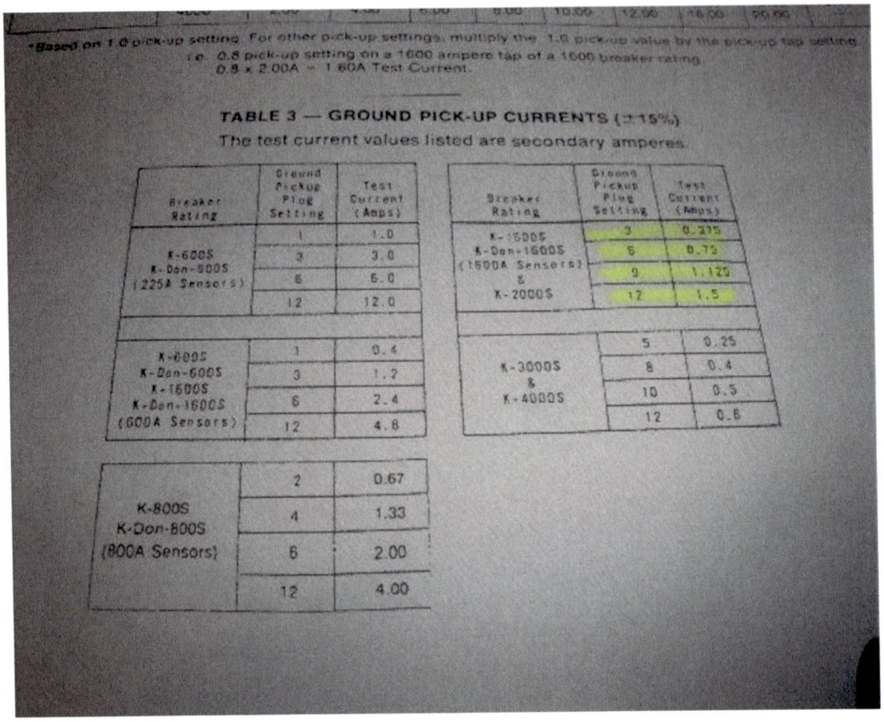

*Based on 1.0 pick-up setting. For other pick-up settings, multiply the 1.0 pick-up value by the pick-up tap setting.
i.e. 0.8 pick-up setting on a 1600 ampere tap of a 1600 breaker rating
0.8 x 2.00A = 1.60A Test Current.

TABLE 3 — GROUND PICK-UP CURRENTS (±15%)

The test current values listed are secondary amperes.

Breaker Rating	Ground Pickup Plug Setting	Test Current (Amps)
K-600S K-Don-600S (225A Sensors)	1	1.0
	3	3.0
	6	6.0
	12	12.0
K-600S K-Don-600S K-1600S K-Don-1600S (600A Sensors)	1	0.4
	3	1.2
	6	2.4
	12	4.8
K-800S K-Don-800S (800A Sensors)	2	0.67
	4	1.33
	6	2.00
	12	4.00

Breaker Rating	Ground Pickup Plug Setting	Test Current (Amps)
K-1600S K-Don-1600S (1600A Sensors) & K-2000S	3	0.375
	6	0.75
	9	1.125
	12	1.5
K-3000S & K-4000S	5	0.25
	8	0.4
	10	0.5
	12	0.6

7. F

GND PICKUP AMPERE TAP	Phase	Calculated Amperes	GO-NO-GO Trip Results
3	B.	.375 Amperes	.250 Amperes
6	B.	.750 Amperes	.510 Amperes
9	B.	1.125 Amperes	1.100 Amperes
12	B.	1.50 Amperes	1.400 Amperes

The table above was created from test results based on the authors actual testing, the lab should experience similar results, the author found 3 tests were sufficient for GO-NO-GO, NOTE: Although the tolerance of ± 15% was not met during the testing. The author determined that the test set was working correctly.

GROUND DELAY TEST

PURPOSE

Verify that pickup occurs within tolerance ± 15%. This test is for **delay pickup** calculations based on 150% of pickup of any of the three pickup settings from the Pick-Up tests. All calculations are based on TABLE 4 –Trip Times for Short-Time Delay, in the **IB-9.1.7.22** manual page 9. For **GO-NO-GO** procedures and long trip results only 3 tests are used.

GO-NO-GO

GF DELAY TEST

Test Procedure

1. Programmer Settings:

   Insure all pins that are not being tested are at their maximum or highest settings.

   Set the Ground Pickup pin to position 12, **See Picture 1.G, page 34.**

   Set pin in GRN Delay Setting to MAX.

2. Position Test Set Controls:

   Set the Function Selector to TIME. **See Picture 2.G, page 34.**

   SET the Phase Selector to desired phase A, B, or C.

   Keep the position of DEFEAT, GROUND, TEST on TEST.

   Set the ammeter scale to 2, **See Picture 3.G, page 34.**

   Preset current to the preset current 1.125 amps or a trip based on **Table 3** of **IB-9.1.7-22** manual, **See Picture 4.G, page 34**.

   Press the Timer Reset button.

   Press the Breaker Reset button.

   Note: The close breaker light should be on.

(Testing Continued) on page 35.

1.G

2.G

3.G

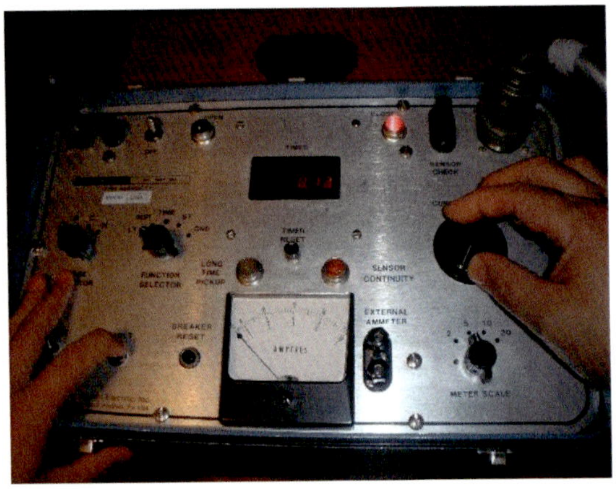

4.G

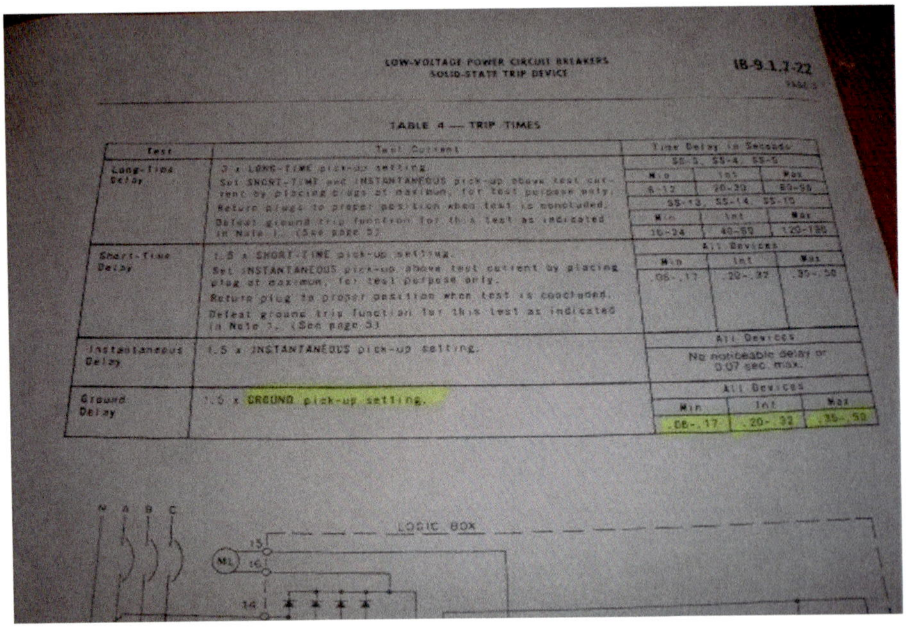

LOW-VOLTAGE POWER CIRCUIT BREAKERS
SOLID-STATE TRIP DEVICE

IB-9.1.7-22

TABLE 4 — TRIP TIMES

Test	Test Current	Time Delay in Seconds		
Long-Time Delay	3 x LONG-TIME pick-up setting. Set SHORT-TIME and INSTANTANEOUS pick-up above test current by placing plugs at maximum, for test purpose only. Return plugs to proper position when test is concluded. Defeat ground trip function for this test as indicated in Note 1. (See page 5)	SS-3, SS-4, SS-5		
		Min	Int	Max
		8-12	20-30	80-95
		SS-13, SS-14, SS-15		
		Min	Int	Max
		16-24	40-60	120-135
Short-Time Delay	1.5 x SHORT-TIME pick-up setting. Set INSTANTANEOUS pick-up above test current by placing plug at maximum, for test purpose only. Return plug to proper position when test is concluded. Defeat ground trip function for this test as indicated in Note 1. (See page 5)	All Devices		
		Min	Int	Max
		.08-.17	.20-.32	.35-.50
Instantaneous Delay	1.5 x INSTANTANEOUS pick-up setting.	All Devices		
		No noticeable delay or 0.07 sec. max.		
Ground Delay	1.5 x GROUND pick-up setting.	All Devices		
		Min	Int	Max
		.08-.17	.20-.32	.35-.50

5.G

3. You can now begin to test:

Hold down the Fault Switch until the Open Lamp lights and the programmer unit trips. The programmer will trip within the time table in Table 4 Trip Times. The settings are based on **Table 4** of **IB-9.1.7-22** manual, **See Picture 5.G, page 34.**

Reset the following;

Timer Reset,

Breaker Reset,

If desired you can change Phase Select to A, B, or C Phase.

Change INT, then to MIN between tests shown on chart below.

On the programmer rotate the magnet on the **GRD target** to the 3 O'clock position and set it back to the 6 O'clock position that will set the target back to black, **See Picture 6.F, page 30.**

GND PICKUP AMPERE TAP	**Phase**	**Time Delay in Seconds**	**GO-NO-GO Trip Results**	**GND DELAY**
12	B.	.08-.17	.11	MIN
12	B.	.20-.32	.28	INT
12	B.	.35-.50	.43	MAX

The table above was created from test results based on the authors actual testing, the lab should experience similar results the author found 3 tests were sufficient for GO-NO-GO procedures.

SENSOR CHECKS

PURPOSE

To verify if the ABB-504 test set is working correctly with the programmer, and that the test set is working correctly for field applications of testing sensors of programmers and complete sensor system checking.

All sensors can be individually checked as follows:

1. Remove the test cable from POWER SHIELD terminal block, **See Picture 1.H, page 37.**
2. Completely remove the test cable from the test set, **See Picture 2.H page 37.**

SIGNAL SENSORS

3. Connect the Sensor Test Lead to the ABB-504 Test Set**, See Picture 3.H page 37.**
4. Connect either lead of the test lead to terminal 5, **See Picture 4.H page 37.**
5. Take the other end of the test lead and touch terminal 8, **See Picture 5.H page 37.**
6. The Sensor Continuity light should light up brightly, **See Picture 6.H page 37.**

   Repeat the terminal checks to terminal 7 and 6.

POWER SUPPLY SENSORS

7. Connect either lead to the test lead to terminal 14.
8. Take the other end of the test lead and touch 11.
9. The Sensor Continuity light will light up **but not as bright** as the previous tests.
10. Repeat the terminal checks to 12 and 13.

SENSOR GROUNDING & MOTOR ALARM TEST

To conduct a SENSOR GROUNDING TEST proper verification can only be performed in the field with a Circuit Breaker Frame. For MOTOR ALARM TESTS the supplied POWER SHIELD programmer is not equipment with these terminals, for **GO-NO-GO** verification the output current has been proved by the previous tests, therefore this test is considered redundant for the ABB-504.

IF Test Results do not conform:

Check first with the **IB-9.1.7-22 Issue B Instruction Bulletin** and then review with the **GO-NO-GO** procedures again.

1.H

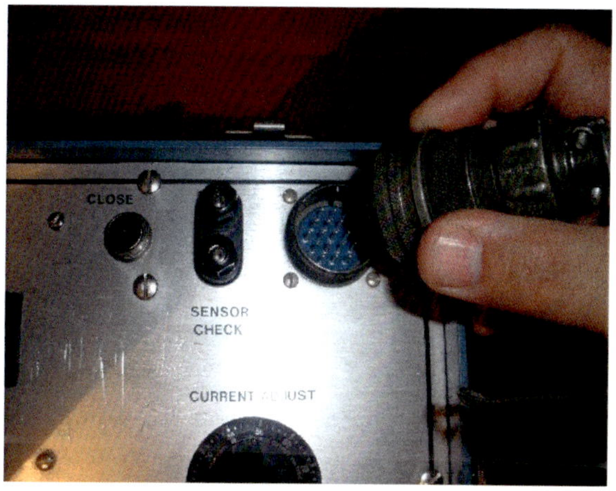

2.H

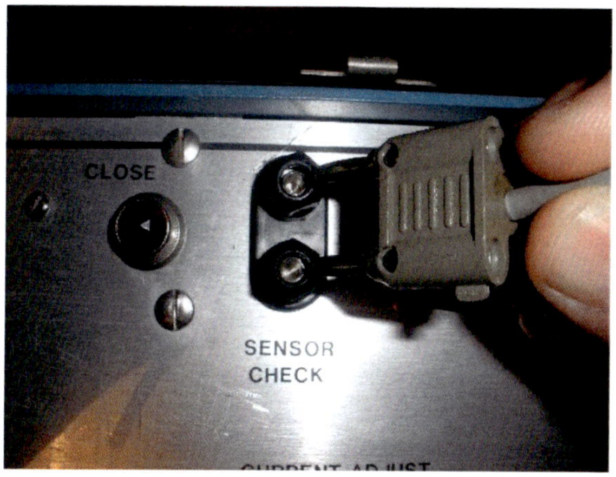

3.H

4.H

5.H

6.H

FINAL NOTE

While creating the **GO-NO-GO** procedures, great care was taken to explain to a technician to verify how to use the ABB-504 power shield test kit while testing a Power Sheild programmer and that it is working properly. The procedures are designed to allow a technician with or without limited experience of testing circuit breakers to accomplish this.

While going over the **IB-9.1.7-22 Issue B Instruction Bulletin** we found the descriptions of the Test Procedures were not always clear with a full explanation or exactly how a test was set up. The Instruction bulletin is provided by the OEM and written so the field technician or engineer performing the test should have working knowledge of all components of the circuit breaker or system being tested.

The **GO-NO-GO** procedures are designed to help an end user navigate testing with the ABB 504-test set, trip unit, and instruction bulletin. With the use of step by step instructions and a pictorial explanation, the goal in this book series is to make it easier for the technician to quickly run through the procedures to make sure they are comfortable using the test set. The **GO-NO-GO** procedures **are not** designed to replace the instruction bulletin, but to complement and assist the end user.

REPAIR, PURCHASE, SPARE PARTS, RENTAL

The ABB 504 Power Shield test set is no longer production however, there is limited availability in the secondary market. Below is where to purchase, spare parts, and rental of the ABB 504.

Madison Testing & Acquisition Services LLC Madison CT, 203-421-9388 www.madisontesting.com	Test equipment sales, spare parts
Solid State Exchange & Repair Co. Denton TX, 940-243-3731 www.solidstaterepair.com	Repair, Sales, Replacement Cables, Repair of trip programmers
Thyritronics Ridgeland MS, 601-856-9798 www.thyritronics.com	Repair, Sales, repair of programmers
Electrical Test Instruments Windsor MD, 410-857-1880 www.electricaltestinstruments.com	Repair, Sales, Calibration
EletroRent Corporation Van Nuys CA, 800-553-2255 Electrorent.com	Test equipment rental, Sales
Protec Equipment Resources Grand Prairie TX, 866-352-5550 Protecequip.com	Test equipment rental, Sales
Advanced Test Equipment San Diego CA, 800-404-2832 Atecorp.com	Test equipment rental, Sales
Intellirent Corporation Roanoke TX, 972-317-0479 www.intellirentco.com	Test equipment rental, Sales
Technical Diagnostic Services Inc Fort Worth TX, 817-465-9494 www.technicaldiagnostics.com	Test equipment rental, Sales

ISBN-13: 978-0692075470
ISBN-10: 069207547X
BISAC: Technology & Engineering / Electricity

GO- NO-GO

END

Made in United States
Orlando, FL
04 November 2026

90198303R00026